SEA LIFE

A Dorling Kindersley Book
Conceived, edited, and designed by DK Direct Limited

Consultant Dr. Malcolm MacGarvin

Editor Sarah Miller
Art Editor Sara Nunan

US Editor B. Alison Weir

Series Editor Sarah Phillips
Series Art Editor Paul Wilkinson

Picture Researcher Paul Snelgrove
Photography Organizer Alison Verity

Production Manager Ian Paton

Editorial Director Jonathan Reed
Design Director Ed Day

First American Edition, 1992
10 9 8 7 6 5 4
Published in the United States by
Dorling Kindersley, Inc., 232 Madison Avenue
New York, New York 10016

Published in Great Britain by Dorling Kindersley Limited.
Distributed by Houghton Mifflin Company, Boston.

Library of Congress Cataloging-in-Publication Data

Sea life / Malcolm MacGarvin, editor. – 1st American ed.
p. cm. – (Picturepedia)
Summary: Introduces a variety of things that live in or near the sea, including crabs, fish, and dolphins.
Includes index.
ISBN 1-56458-140-3
1. Marine biology – Juvenile literature. [1. Marine animals.]
I. MacGarvin. Malcolm. II. Series.
QH91.16.S43 1992
574.92 – dc20 92-52836
CIP
AC

Reproduced by Colourscan, Singapore
Printed and bound in Italy by Graphicom

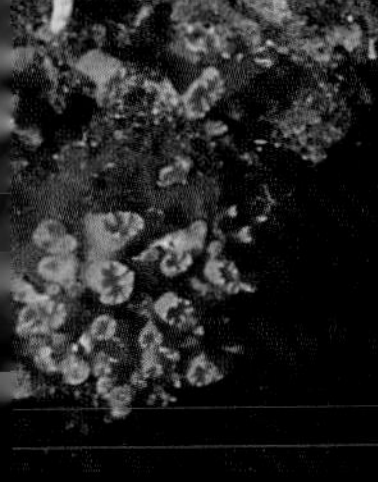

SEA LIFE

DORLING KINDERSLEY, INC.
LONDON • NEW YORK • STUTTGART

CONTENTS

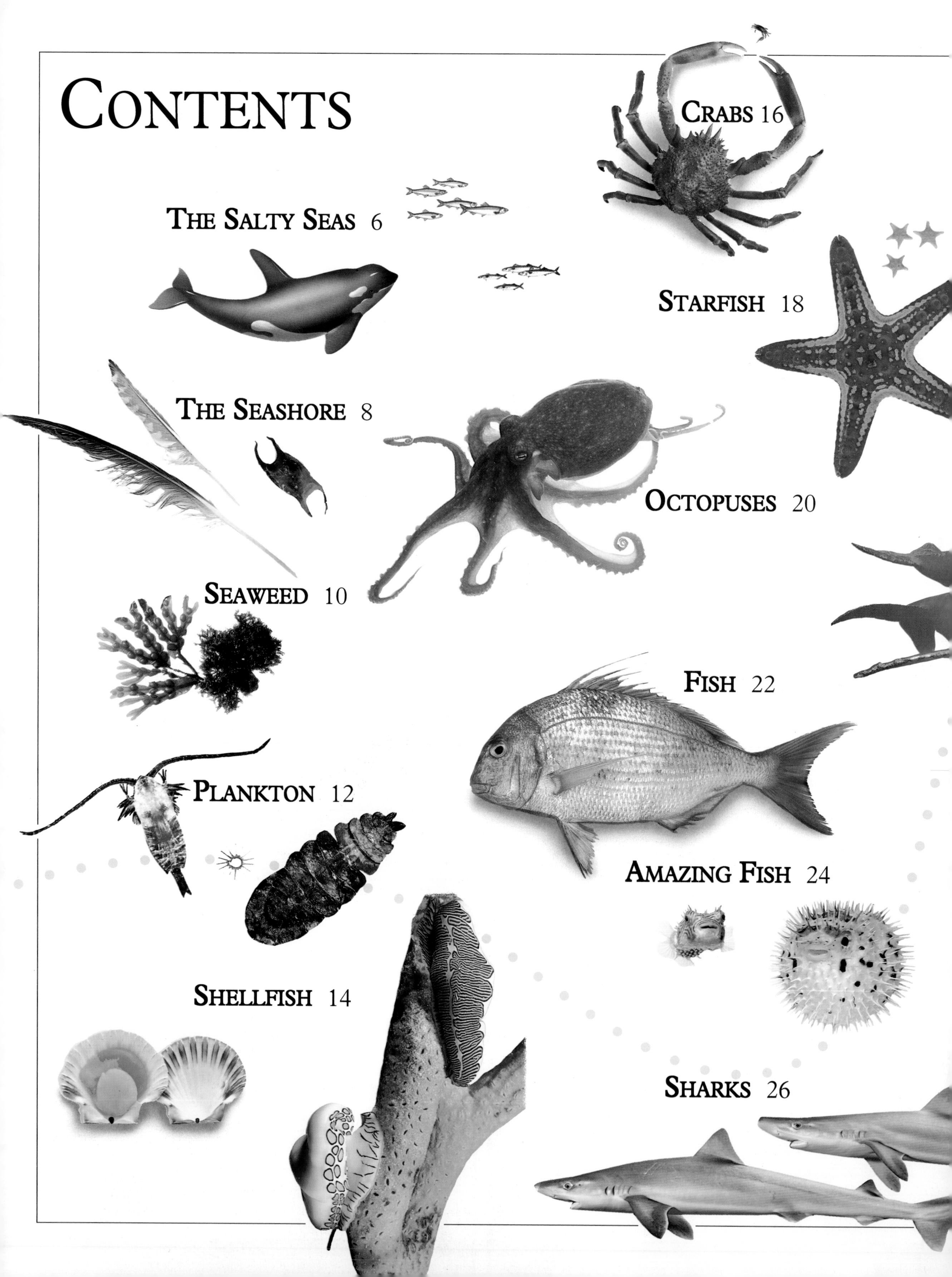

THE SALTY SEAS

Almost three quarters of the Earth is covered by oceans and seas. These billions of tons of salty water are home to silent sharks and playful dolphins, huge whales and tiny fish. There are even beautiful underwater gardens of coral. Many parts of the ocean have never been visited by people, so there may be even more amazing plants and animals in this marvelous, mysterious, watery world.

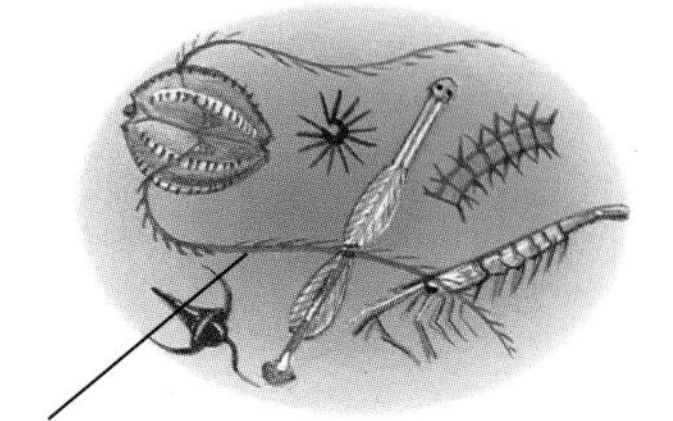

Tiny plants and animals, called plankton.

Life Beneath the Waves
Thousands and thousands of plants and animals live in oceans all over the world.

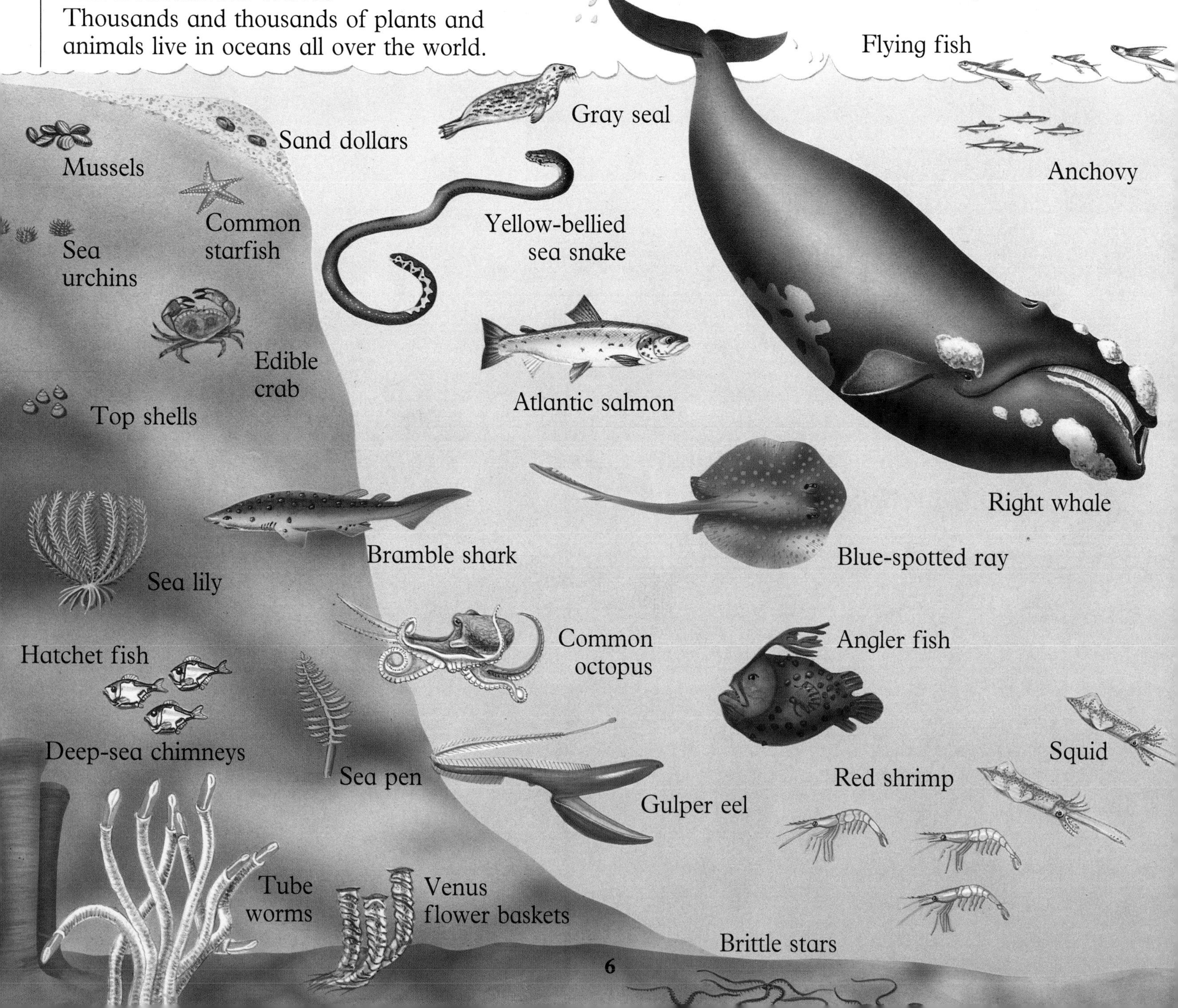

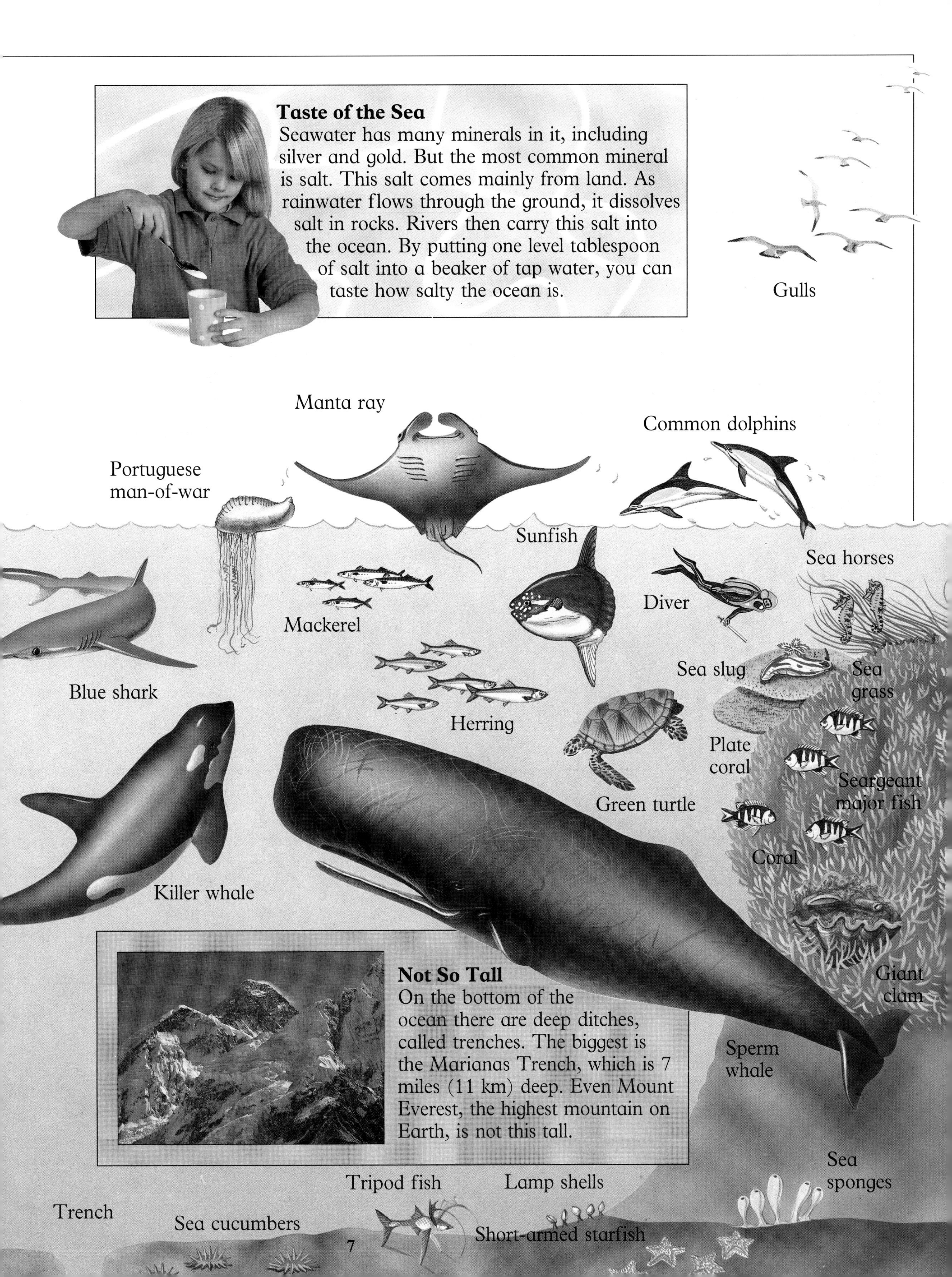

Taste of the Sea

Seawater has many minerals in it, including silver and gold. But the most common mineral is salt. This salt comes mainly from land. As rainwater flows through the ground, it dissolves salt in rocks. Rivers then carry this salt into the ocean. By putting one level tablespoon of salt into a beaker of tap water, you can taste how salty the ocean is.

Not So Tall

On the bottom of the ocean there are deep ditches, called trenches. The biggest is the Marianas Trench, which is 7 miles (11 km) deep. Even Mount Everest, the highest mountain on Earth, is not this tall.

THE SEASHORE

When you visit the ocean, the beach is sometimes very wide. At other times, it can't be seen at all. This is because every day, twice a day, the sea moves up and down the shore. These movements are called tides. Plants and animals that live here can survive under water, and also when the tide is out and the land is dry.

Scallop shell

Gull feather

Miniature World
When the sea flows down the beach, water is trapped in hollows among the rocks. These little lakes are called tide pools. Many creatures live in these wave-swept, watery worlds.

This starfish is climbing on mussels.

When the tide comes in, waves crash into the tide pool. This limpet will not be knocked off its rock – it glues itself to the rock with sticky slime and holds on tight!

Beadlet anemone

Is it a pebble? No, it's a crab!

Tube worms make these long, chalk tubes to hide in.

Sea urchin

Broken limpet shell

Treasure Hunting
The best time to look for "treasures" that have been washed up by the sea is at low tide. But be careful. The sea can come back quickly, and you could find yourself surrounded by water!

Egg cases of the common whelk

If you dig in wet sand, you may find shellfish. They live underground when the tide is out.

Part of a crab's shell

Gull track

Dogfish egg case, or "mermaid's purse"

Weird Eggs
All sorts of egg cases are washed ashore. You may find an odd-shaped shark egg case.

This small pile of sand is called a wormcast. It is the waste of a sand-eating lugworm.

This shell comes from the inside of a cuttlefish.

Put seaweed and rocks back where you find them, because crabs and fish use them to hide in.

Make Your Own Viewer
If you use a tide pool viewer, you will be able to see the animals in a tide pool much more clearly. You will need:

- *1 foot (30 cm) of plastic pipe*
- *Lucite sheet*
- *waterproof tape*

Place one end of the pipe onto the Lucite and draw around it. Ask an adult to cut this circle out, then tape it to the end of the pipe. Now dip the viewer into a tide pool and see an amazing mini-world!

Red seaweed

Top shells eat seaweed, but this one is stuck to a dog whelk!

What a Difference!
When anemones are covered by the sea, they look like underwater flowers. But when the tide is out, they have to pull in all their tentacles or they will dry up. This makes them look like blobs of jelly!

SEAWEED

Seaweeds are plants that live in the sea. Like all plants, they use sunlight to make their food, so they grow only in water where there is plenty of light. But unlike most land-living plants, seaweeds have no flowers or roots.

Ready for Bed
Sea otters wrap themselves in seaweed before they rest. This keeps them from drifting out to sea while asleep.

Home Sweet Home
Many animals, such as starfish, live on or around seaweed.

Air inside these bubbles helps seaweed float.

Most kinds of seaweed are slimy. This slime keeps the fronds wet when the tide is out.

Colors on the Beach
There are green, red, and brown seaweeds. Green seaweeds, such as wracks, need lots of light, so they grow nearest the shore. Oarweeds, and other brown seaweeds, can live in slightly deeper, darker water.

Kelp-eaters
Purple urchins eat kelp. If there are enough urchins, they can eat entire forests of giant kelp.

Channeled wrack

High tide – the sea rarely goes above this point.

Serrated wrack

Bladder wrack

Sea oak

This rock is covered in a thin layer of red seaweed.

Sea lettuce

Lowtide – below this point, the land is always under water.

Red feather seaweed

Thongweed

Oarweed

This leathery strip is called a frond. It bends, but does not snap, when waves crash into it.

Sugar kelp

Underwater Forests
Giant kelp is the largest plant in the ocean. It can grow to a height of 100 yards (109 m) – as tall as a 15-story building!

Thongweed

Dulse

Bladder wrack

This red seaweed is called carrageen, or Irish moss.

The tip of this frond has been ripped by the waves.

Seaweeds have finger-like holdfasts that cling to rocks.

PLANKTON

Seawater is full of billions of very tiny plants and animals called plankton. Most types of plankton are less than one millimeter long, but without them, very little else could live in the sea! They are the most important food for many fish, whales, and even birds. Their tiny size means that they are not strong enough to swim against the water's current, so they can only drift around.

This curved feeler is covered in fine sensory hairs that help the copepod find food.

Animal plankton can be more than 100 times bigger than the plants they eat.

Micro Metropolis
Water fleas, called copepods, are a type of animal plankton. They are so small that more than one million could fit into a bucket of seawater.

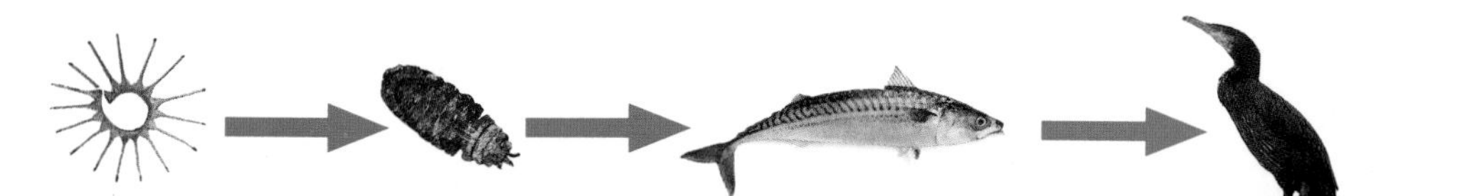

Plant plankton...eaten by...animal plankton...eaten by...fish...eaten by...birds.

Food for Everyone
Birds, sea turtles, and seals all eat other animals, such as fish and shellfish, that eat plankton. They would all starve if there were no plankton in the ocean.

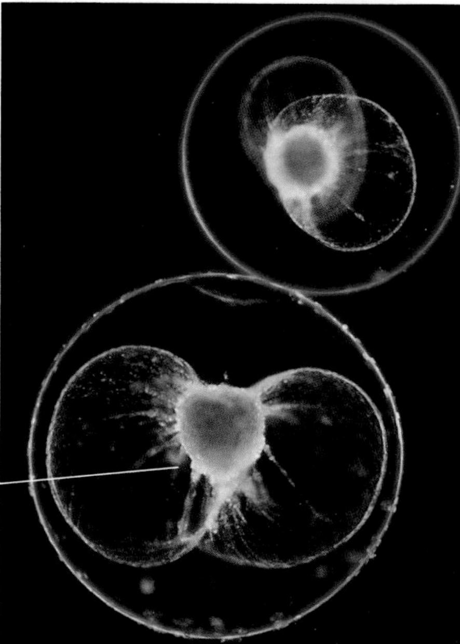

The dark water shows through the colorless parts of this plant plankton.

Sun-lovers
Like all plants, plant plankton use sunlight to make their food.

Part-time Plankton
Some animals begin their lives as tiny plankton, but then they grow much, much bigger.

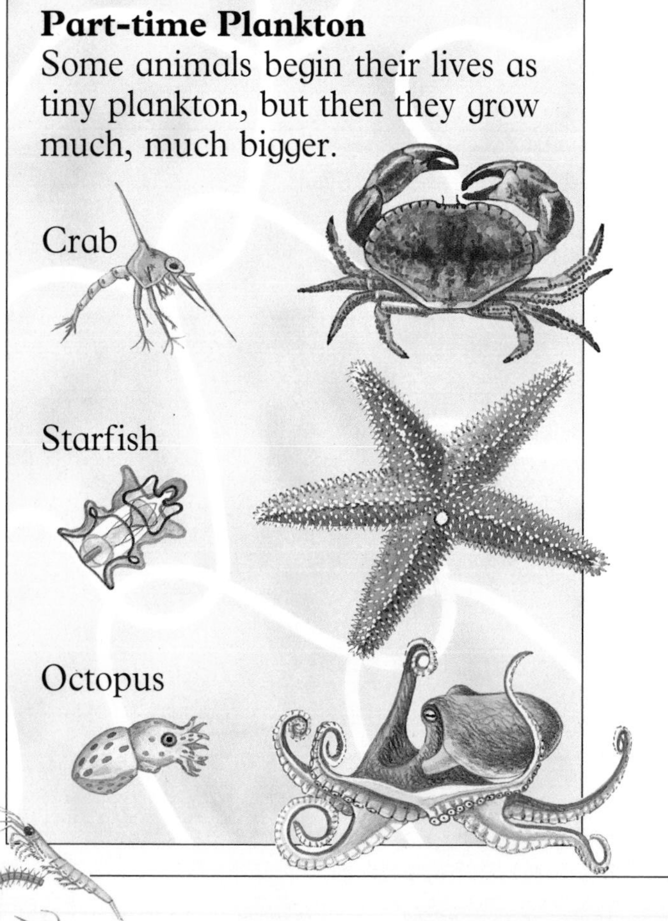

There are many different kinds, or species, of copepods. This male is flashing bright blue to attract a female.

By waving these bristles, the copepod pushes water over its tiny gills and also traps plant plankton.

Countless Copepods
Copepods are the most common animals in the sea.

Eye

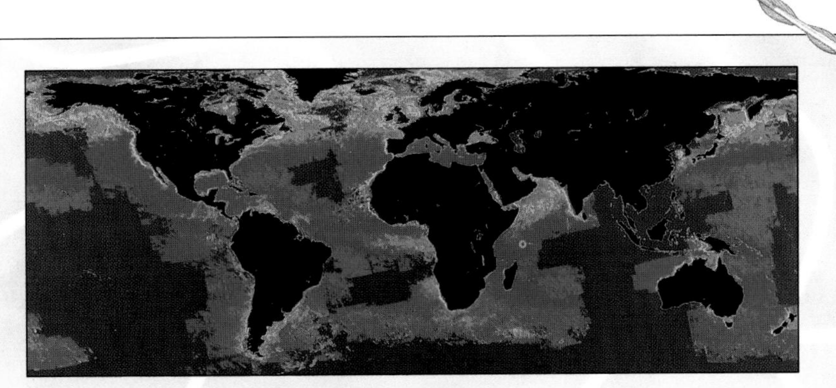

Soupy Seas!
Plant plankton live near the surface of the ocean where there is lots of light. They also prefer cooler water because it has more minerals in it. Animal plankton eat plant plankton, so they live in the same places. This map of the world shows where they both live. The red, yellow, and green areas have more plankton than the blue, purple, and pink areas. Plankton has not been counted in the gray parts.

At night, copepods drift nearer to the surface and hunt for food. During the day, they sink deeper down to hide from hungry birds and fish.

Copepods can't swim against the current, but they can move by using their feelers as oars.

Tail

Zooplankton (Animals)

Sea gooseberry

Krill

Arrow worm

Phytoplankton (Plants)

Lots of plant plankton, or phytoplankton, have joined together to make this chain. The chain is too big to be eaten by many zooplankton.

SHELLFISH

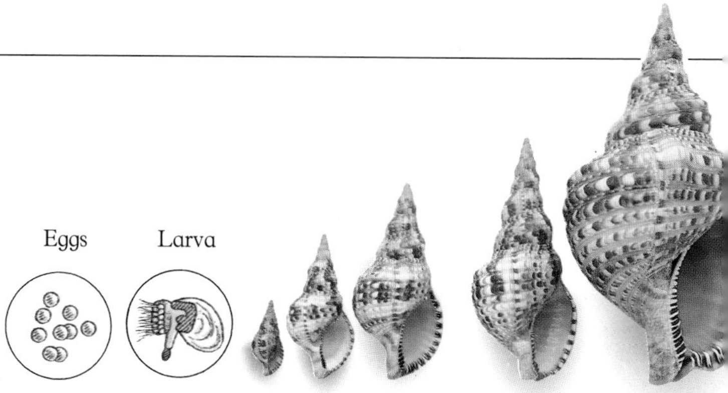

The shells that you find on a beach are the empty homes of small animals called shellfish. These soft-bodied creatures need hard shells to protect them from starfish, crabs, fish, and even birds. The most common types of shellfish are gastropods and bivalves. Gastropods are underwater snails. They grow coiled shells and slide around on a slimy foot. Bivalves have two, flatter shells that cover their whole bodies.

Eggs

Larva

Growing Up
Unlike crabs, shellfish never need new shells. Baby shellfish, or larvae grow a tiny shell that keeps on getting bigger.

The Animal Inside
Bivalves, gastropods, and all other shellfish have soft bodies and no backbones. They are mollusks, just like octopuses!

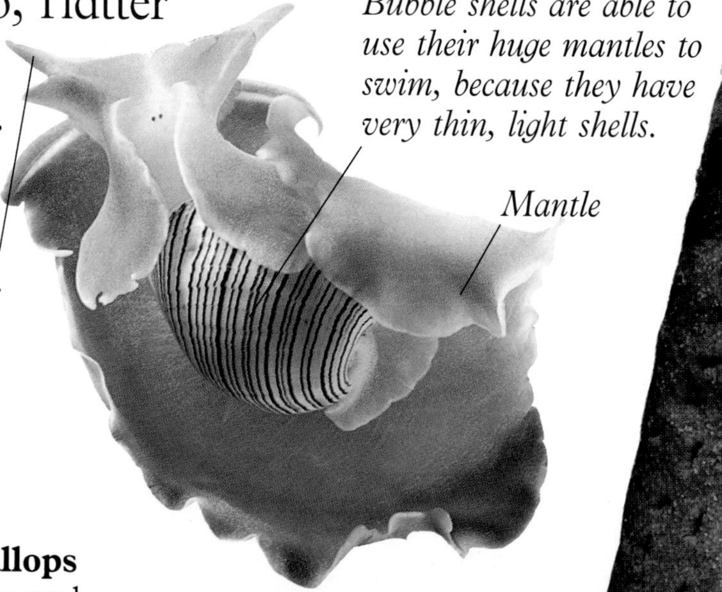

Bubble shells are able to use their huge mantles to swim, because they have very thin, light shells.

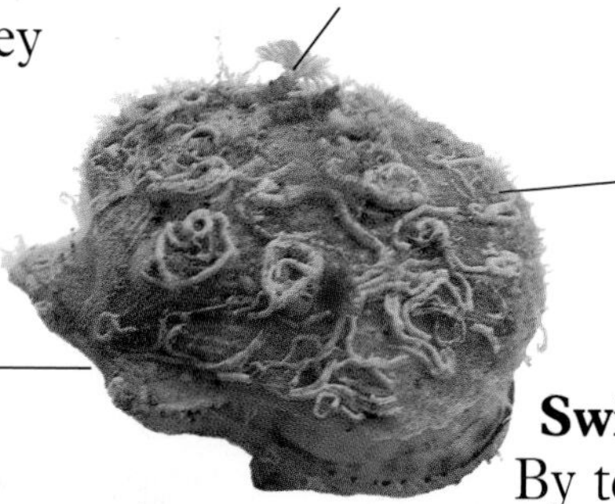

Mantle

These feelers scan the water for food.

Worm

Water in

Water out

Swimming Scallops
By taking in water and then shooting it out its back end, scallops propel themselves through the sea.

Eye

Buried Alive
Many bivalves spend most of their lives buried in sand. They dig a hole with their foot, then poke tubes, or siphons, into the sea. The siphons take water into the gills and catch food.

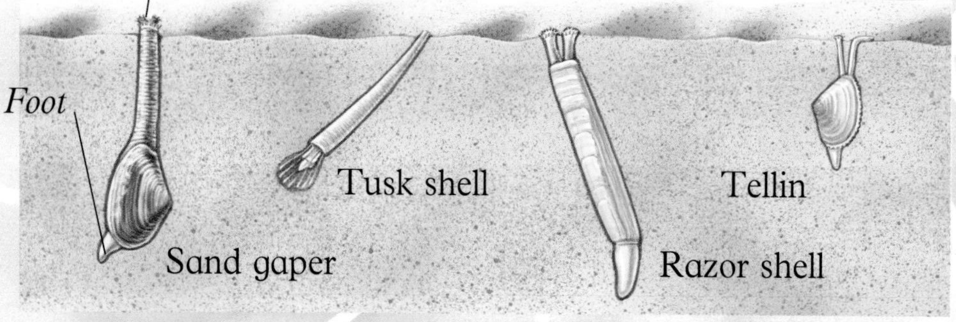

Siphon

Foot

Tusk shell

Tellin

Sand gaper

Razor shell

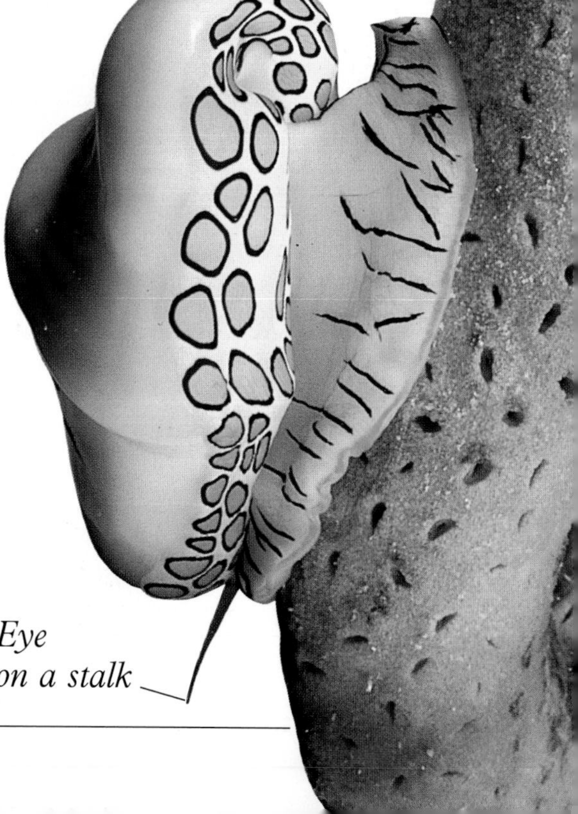

Eye on a stalk

Giant Clam

The biggest shellfish in the world is the giant clam. This huge bivalve lives in warm waters and is more than a yard (1 m) wide. You could easily fit inside its two big shells, but you wouldn't get trapped. Giant clams can only close their shells very slowly, giving you plenty of time to escape.

The Shellfish Family

Gastropod

Chiton

Bivalve

Tentacles feeling for plankton

Tusk shell

Big foot

This soft, brightly colored skin is called a mantle. It can wrap right around the shell. As it slides over the shell, it smooths away scratches on the surface.

White shell

When it is frightened, this fingerprint flamingo tongue shellfish will hide inside its shell.

Coral

Coral-suckers

These two belly button cowries are using their rough tongues to pull off tiny coral animals. Other shellfish eat seaweed and even fish!

Spotted foot

This cowrie's mantle is spotted like the coral it is sitting on. This helps hide it from hungry starfish.

CRABS

What lives in the sea or on land, can be any color, has its eyes on stalks, swims and walks sideways, and carries its own house? Would you have guessed a crab? Crabs live in all parts of the sea, from the very deepest oceans to wave-swept shores.

Crabby Face

Antennae

Eye

Mouth

Crabs use their rear four pairs of legs to scuttle sideways.

Its strong claws are used for fighting and also for tearing apart fish, shellfish, and plants to eat.

Under this spine, there are two feelers, called antennae. The crab uses tiny hairs along the antennae to touch, smell, and taste.

Its shell is often called a carapace.

When crabs get too big for their shells, they split them open and shed them. Underneath the old suit of armor, there is a new, soft shell. This can take three days to harden.

Crabs can grow new legs if they are torn off.

This spiny spider crab is protected from most of its enemies by its shell.

Crabs breathe through five pairs of gills. These are inside the shell, near the top of each leg.

Joint

Crab-eaters

Crabs make a tasty meal for fish, birds, octopuses, seals, and people!

Shore crab

Stony-shelled crab

This crab is so ugly that it is called the horrid crab!

The claws of the common lobster are strong enough to snap off your finger!

Anemone *Hermit crab* *Feeler*

Close Quarters
A hermit crab does not have a hard shell to protect it. Instead, it lives in old snail shells. A sea anemone has also made its home on this shell. It eats food that the crab drops and stings enemies that come too near their home.

Crusty Crustaceans
Crabs are crustaceans. This means that they have a crust, or a shell, and at least five pairs of legs. There are thousands of kinds of crustaceans. Here are a few common ones:

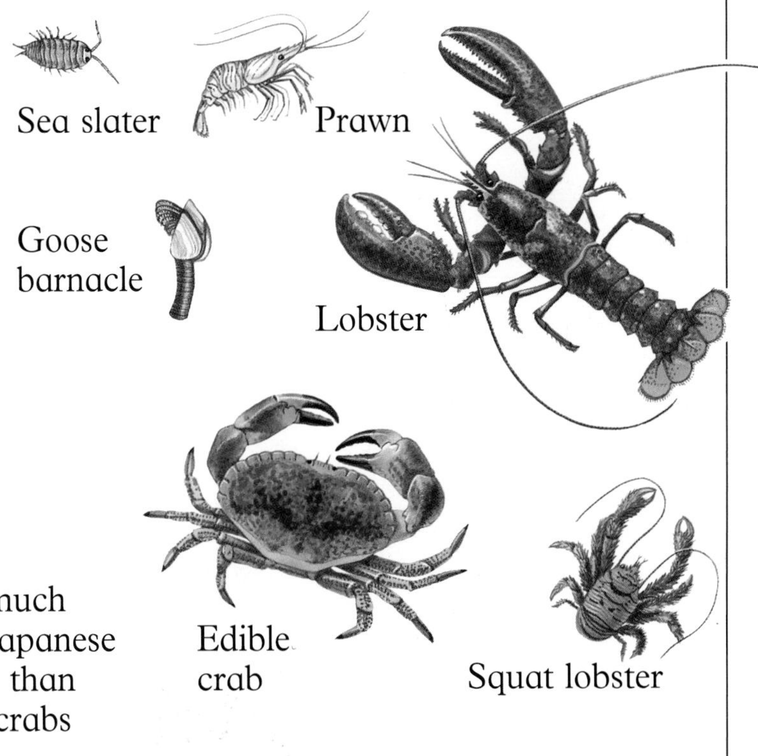

Pea crab, actual size

Little and Large
Most crabs are about 6 inches (15 cm) wide, but a few are much bigger ... or much smaller. Japanese spider crabs grow to more than 13 feet (4 m), while pea crabs are the size of a pea!

Japanese spider crab claw, half actual size

STARFISH

Starfish are star-shaped, but they are not fish – they are echinoderms. This means that they have spiny skin. They cannot swim, but they are very good at crawling! They can walk up seaweed fronds and climb down the sides of rocks. Even in the deepest, darkest parts of the sea, there are starfish creeping around.

This crimson-knobbed starfish, like most starfish, measures less than eight inches (20 cm) from tip to tip. But some species are as wide as a small car!

Starfish don't have eyes. Instead, they have eyespots on the tips of their arms. These special cells cannot see shapes, but they can tell whether it is light or dark.

These bumps are actually spines.

A starfish's arms are very flexible. This is because its skeleton is made up of lots of tiny spines that can move in any direction.

Central disc

Starfish breathe through their feet and also through tiny tubes that are found all over their bodies.

Burrowing starfish

Cushion star

Common starfish

Stomach This
Starfish eat shellfish. When a starfish finds a clam, it pulls open its shell with its tiny tube feet, pushes its whole stomach inside the shell, and then slowly digests the clam.

Common sea urchin

Slate-pencil sea urchin

Sea Porcupines
Sea urchins are close relatives of starfish. Their long spines, which are sometimes poisonous, make them look like porcupines.

Goosefoot starfish

Some starfish have more than five arms. This spiny sun star has twelve!

Newly Armed
This starfish is growing two new arms. The old ones were bitten off by a fish! As long as the central disc and one arm is left, the starfish will survive.

New arm

Spiny Species
There are more than 6,000 different species of echinoderms, but only five main groups. These are:

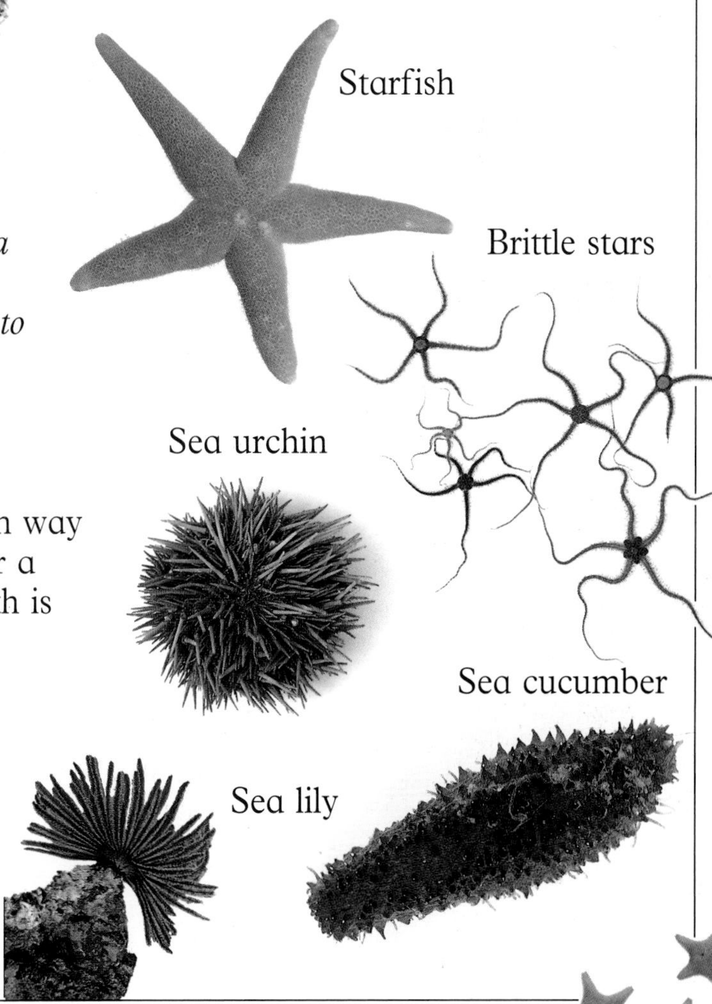

Tube foot

Each of these feet has a sucker on the end that helps the starfish stick to rocks and catch food.

Flip Side
It is easy to tell which way is up for a starfish or a sea urchin – its mouth is always underneath.

Most starfish have five arms.

If a starfish flips over, it uses its arms to right itself.

Mouth

OCTOPUSES

Bunch of Eggs
Inside each of these soft-shelled eggs, there is a baby octopus.

Did you know that octopuses are related to snails? But unlike most other mollusks, they don't have shells to protect them. Instead, these eight-armed animals squeeze their soft bodies into small cracks or holes in rocks. Once they are safely hidden, it is very hard for conger eels, sharks, seals, and people to find and eat them.

Octopuses can see shapes and colors very well with their large eyes.

Gone Fishing
Octopuses hunt for their food. They pounce on fish, starfish, and crabs. Some have webs between their arms that help them net even more animals.

Tentacle

Web

This common octopus is about four inches (10 cm) wide. The largest octopus ever found was more than 29 feet (9 m) from tip to tip!

An octopus can shoot ink, called sepia, out of its siphon. This black cloud hangs in the water and hides the octopus from its enemies.

Siphon

Jet-propelled
If an octopus is frightened, it does not crawl slowly away – it jets off! By forcing water out through its siphon, it can shoot through the water.

Suckers

Rows of super-strong suckers help octopuses hang on to rocks, grab food, and pull themselves along the ocean floor.

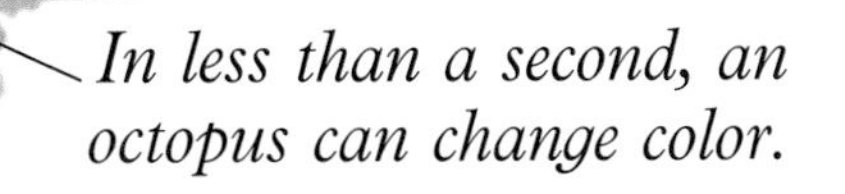

In less than a second, an octopus can change color.

The soft body of an octopus is like a big bag of skin. Water flows into this stretchy bag, passes over the gills, and then escapes through a special funnel, called a siphon.

Sucker

The Cephalopod Family

Octopuses and their relatives are known as cephalopods. This means they are mollusks that live in the ocean and have tentacles.

Cuttlefish

Vampire squid

Squid

Nautilus

Octopus

Battle of the Giants

There are stories of huge octopuses swallowing people whole, but this doesn't actually happen. These myths may be based on giant squid, which can be 65 feet (20 m) long and are close relatives of octopuses. Deep down in the sea, giant squid are believed to battle against sperm whales.

Sperm whale

Giant squid

FISH

All living things need a gas called oxygen to breathe. You cannot see it, but it is found in air and water. You use your lungs to breathe in air. If you swim under water, you either have to hold your breath or use a snorkel. Fish don't have to do this. They can take their oxygen straight out of the water.

Fish Eggs
Most fish lay jellylike eggs. Some guard their eggs until they hatch. Others, like cod, just squirt millions of tiny eggs into the sea. This is called spawning.

Dorsal fin

Fish do not need eyelids – seawater keeps their eyes wet and free of dirt.

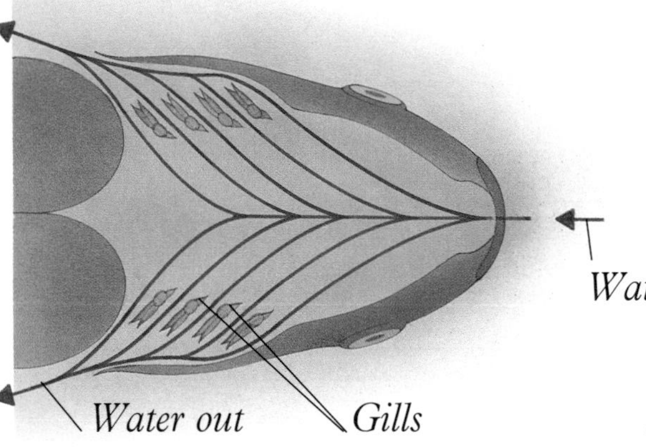

Pectoral fin

Breathing Under Water
Water flows into a fish's mouth, over flaps called gills, and out through the gill openings. The gills take the oxygen out of the water.

There are four gills behind this opening.

Bony fish have a "balloon" inside them called a swim bladder. The air in this organ helps fish stay afloat.

Its pectoral and pelvic fins help a fish move up, down, left, or right.

Pelvic fin

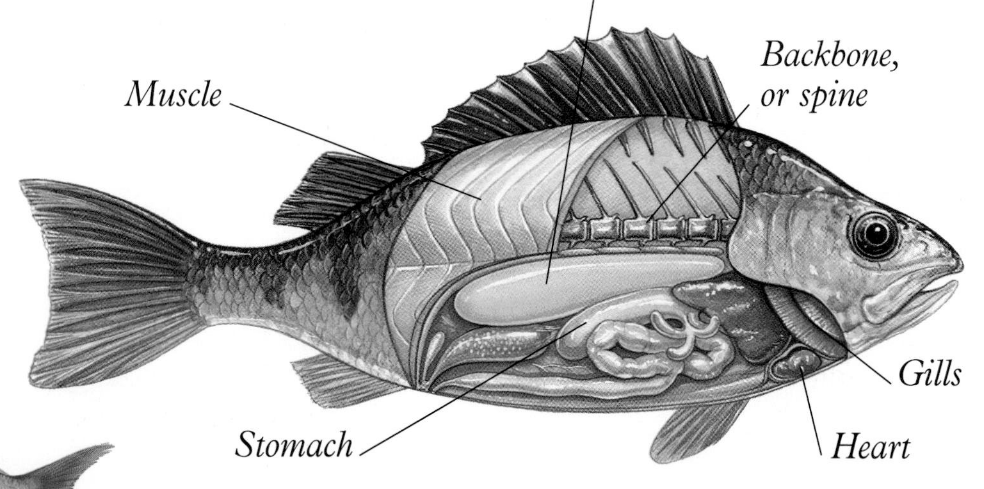

Inside Story
Most of the important parts of a bony fish are in the lower half of its body. The top half is full of muscles that move its tail.

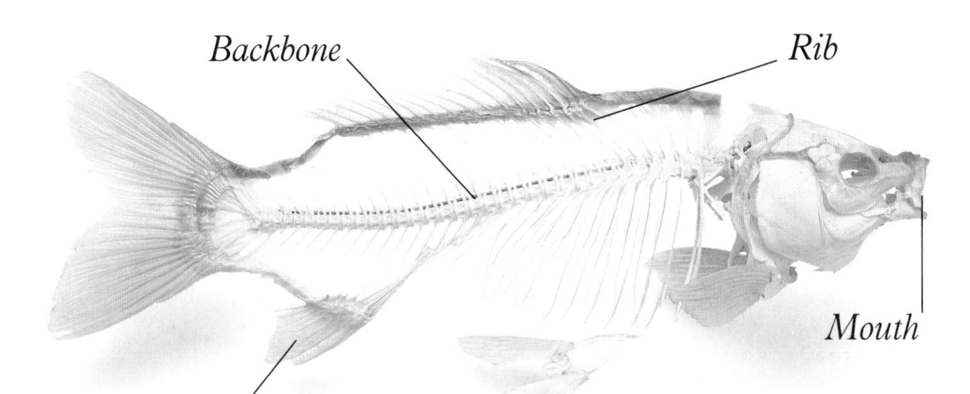

A Bony Back

The bones inside an animal are known as a skeleton. Fish were the first animals to develop a backbone.

This sea bream is covered in thin scales that overlap like the shingles on a roof.

Fin ray

Fish move their tails, or caudal fins, from side to side to power themselves through the sea.

Anal fin

Its dorsal and anal fins stop a fish from rolling over in the water.

Fish have a long, thin tube just under the surface of their skin, called a lateral line. It has little lumps of jelly next to it that wobble when the water moves. Fish can feel this – so they know if something is moving near them.

Fish Families

There are three types of fish: those with hard skeletons made of bone, those with skeletons made of rubbery tissue called cartilage, and those that do not have jaws.

Bony Fish

Mackerel

Sea horse

Flounder

Cartilaginous Fish

Spotted leopard shark

Jawless Fish

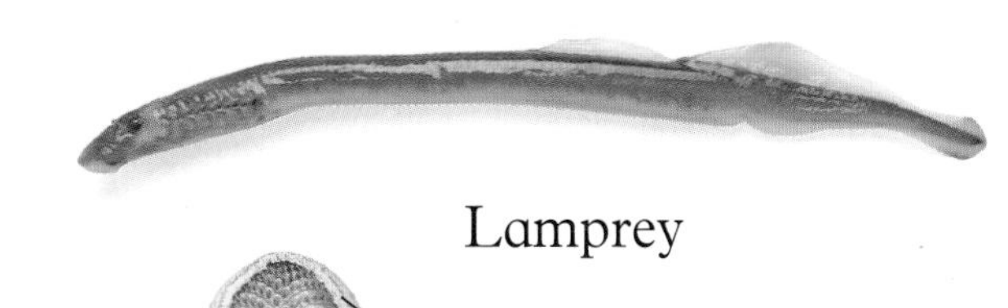

Lamprey

A lamprey has a sucker instead of a mouth.

AMAZING FISH

During the 450 million years that fish have swum in the seas, they have developed some amazing shapes to help them catch their food or hide from their enemies. Many are streamlined – they have pointed heads and smooth bodies. This helps them

speed through the water. Others have sharp spines, big mouths, or frilly fins to help them survive.

The Smallest Fish
Some dwarf gobies are less than a half-inch (13 mm) long. One could easily fit on the tip of your little finger.

The Biggest Fish
A whale shark is heavier than three elephants and longer than five family cars!

Diver

Big mouth

Fish Food
Different fish eat different things, which they catch in all sorts of ways. Some have pointed teeth and chase their food through the ocean. Others have much stranger ways of catching their food.

A slingjaw fish can shoot out its whole jaw to grab small fish and shrimp.

This parrotfish eats coral that is as hard as rock! It has tough lips to break off the coral and flat teeth to crunch it into little pieces.

Beautiful butterfly fish poke their long snouts into cracks to suck out worms.

Rainbow Colors
Just like players on a baseball team, some fish wear brightly colored "uniforms" so they can find one another in a crowd.

This odd-shaped animal is a ghost pipefish. Big hungry fish think it is just a piece of seaweed and so they do not try to eat it.

Mouth

Spine

Eye

A prickly blowfish scares away its enemies by gulping down water and blowing itself up like a balloon!

Weird and Wonderful
Some fish are not fish-shaped at all! They use their strange shapes to hide or to frighten bigger fish away.

The long, fanlike fins and the striped body of this lionfish warn other animals that it has poisonous spines.

Back from the Dead
In 1938, a fisherman caught a fish, called a coelacanth, that had flipperlike fins. This type of fish was thought to have died out more than 90 million years ago.

SHARKS

Sharks are some of the best hunters in the ocean. With their razor-sharp teeth and huge jaws, they can tear up seals, fish and even wooden boats! Tiger sharks and great white sharks sometimes attack people, but most sharks are frightened by people and avoid them.

See how big this jaw is!

Sharks are like a swimming nose. They can smell injured animals and other food that is hundreds of yards away.

Teeth as Sharp as Knives
A shark's teeth can cut through skin and crunch up bones, but they soon get blunt. Each tooth only lasts for a few weeks, then it falls out and is replaced by a new one. Basking sharks eat plankton, so they don't have any teeth!

Dorsal fin

All living animals produce a small amount of electricity. You cannot see it – but sharks can. Little dimples on their heads work like television antennae to tell them where other animals are hiding.

Five gill slits on each side of the head let water into the shark's gills.

Pectoral fin

Super Swimmer
A shark swims by bending from side to side. First it moves its head, then its body, and last of all its long tail. As this wave travels down its body, it pushes the shark through the sea.

Shark Attack!
Just before they bite, sharks bend their noses up and thrust their teeth forward.

Great white shark

Sandtiger shark

Blue shark

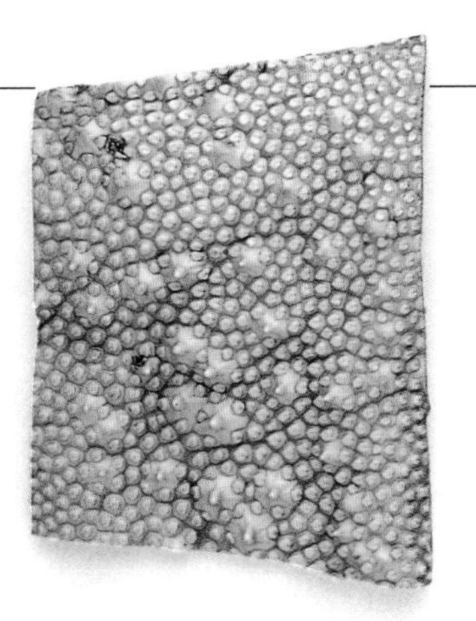

Close-up of shark skin

Watch Your Hands
Shark skin feels like rough sandpaper – it is covered in tiny "teeth."

Tail, or caudal, fin

Anal fin

Like most sharks, this tope shark has a second dorsal fin.

The top of a shark's tail is larger than the bottom. This odd shape helps to raise the shark in the water when it is swimming.

Pelvic fin

Shark Shapes
Not all sharks look like the ones you see in movies: some are tiny and others have odd shapes.

Cookiecutter shark

Hammerhead shark

Prickly dogfish

Wobbegong

Great white shark

Some great white sharks are 26 feet (8 m) long.

Sink or Swim
Sharks, like many fish, are heavier than water, so they should sink to the bottom of the sea. Bony fish have inflatable swim bladders to stop this from happening, but sharks have oily livers instead. The oil helps them float because it is lighter than water. There is enough oil inside a basking shark's liver to fill five big buckets!

Liver

Porbeagle shark

Graceful shark

Pelagic thresher shark

Oceanic whitetip shark

RAYS

Rays are some of the most common fish in the ocean. A few look like their relatives, the sharks, but most of them are diamond-shaped and flat. Some of these shy, gentle animals can be glimpsed gliding through the sea, but others spend their time lying on the ocean floor.

These sharp spines protect the ray. Stingrays also have one very large poisonous spine that they use to stab their enemies.

Thornback ray

Like sharks, rays have skeletons made of cartilage.

Gill slit

Mouth

Small pelvic fin

Upside Down
On the underside of a ray, there are ten gill slits and a mouth that is full of flat teeth.

All rays have a tiny hole, called a spiracle, behind their eyes. This hole lets clean water enter the gills when the gill slits are blocked. This happens when a ray lies on the bottom of the ocean.

Mighty Manta
Most rays are less than two yards (2 m) wide but a manta ray can grow to more than eight yards (8 m) in width. Luckily for divers, it only eats plankton and small fish.

By flapping their fins, rays disturb the sand and uncover crabs – a tasty snack for a ray.

Underwater Flying
A ray is one of the most graceful swimmers in the sea. Flapping its huge fins, it "flies" like a giant underwater bird.

Shocking Ray
Electric rays have special muscles in their bodies that act like batteries. Once they have pounced on a fish, they kill it by using those muscles to produce more than 200 volts of electricity. The shocks are powerful enough to stun people who accidentally step on these rays.

These swirly patterns help hide the ray when it lies on sand.

Like all rays, this undulate ray's large pectoral fins are joined to its head.

Spiracle

Eye

Rays have a very good sense of smell.

Peek-a-boo
When rays hide in the sand, their large, bulging eyes stick out. They have to stay on the look-out for food and also for sharks that like to eat rays.

Ray Shapes
There are over 450 different types of rays, half of which are skates.

Sawfish

Guitarfish

Skate

Eagle ray

Stingray

Coral Reefs

Coral reefs are like beautiful underwater gardens. They are made up of millions of tentacled animals, called corals, which live in huge groups, or colonies. Each coral lives in a tiny, chalky cup that it builds itself. When it dies, its white stone home remains, and another coral then builds its home on top of this cup. Over many years, piles of cups build into massive mounds called reefs. The caves, valleys, and cliffs in these reefs are home to deadly eels, fantastic fish, and shimmery sea slugs!

A "Rock" that Grows
Corals grow at about the same speed as your fingernails. This piece must have taken a long time to get so big!

Clowning Around
These clownfish safely swim and hide among the sea anemones' tentacles. The clownfish is immune to its "friend's" poisonous sting.

Corals have plants, called algae, living inside them. Algae make food for corals.

Birth of an Island!
When coral reefs grow so tall that they stick out of the sea, they turn into a special kind of island called a key.

Coral reef

Baby corals are born in two ways: some hatch from eggs that float in the ocean; others grow out of their parent, like buds on a plant.

The Living Animal
Corals that build cups are called hard corals. Those that do not are called soft corals. Most of these plankton-eating animals are only a little bigger than the head of a pin!

The Coral Family

Waving in the Water
During the day, hard corals hide inside their stony homes. But at night, they wave their tiny, stinging tentacles in the water to trap small bits of floating food.

Soft corals can be almost any color – red, blue, or yellow. But they do not make cups, so they can't build a reef.

Hard coral

HIDE AND SEEK

The ocean can be a dangerous place to live. It is full of hungry sharks, seals, and killer whales. So how do small, defenseless animals avoid being eaten? Some can swim swiftly away from danger. Others change the way they look and try to hide. A few scare their enemies away by pretending to be tough! In this fish-eat-fish world, all sorts of clever tricks are needed to survive.

Shell

Fancy Dress
These sea urchins have stuck pebbles and bits of shell to their spiky bodies. Now they look like piles of rocks and won't be eaten.

What is Camouflage?
This boy is wearing clothes that are a similar color to the background. This makes him harder to see – he is camouflaged. The girl's red clothing is much more visible.

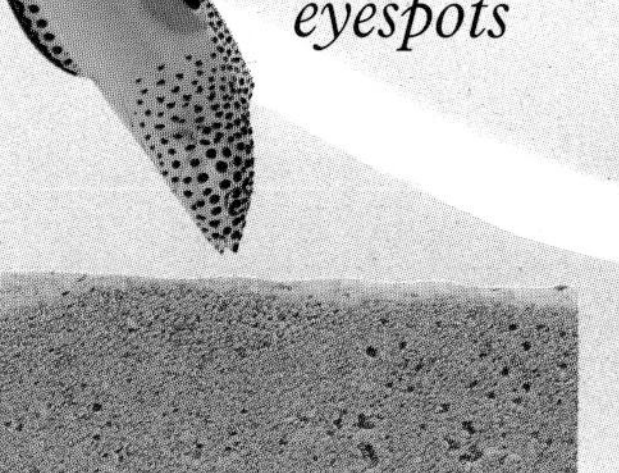

False eyespots

Quick, Hide!
If you can't run, you'd better hide. That's what this small twinspot wrasse is doing.

We Don't Taste Nice!
Some animals want to be seen. Their brightly colored bodies warn other animals that they are poisonous.

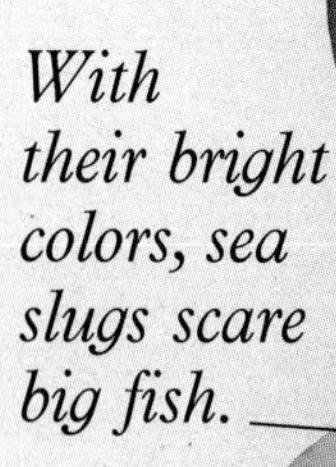

Gills

With their bright colors, sea slugs scare big fish.

The blue-ringed octopus has blue circles on its yellowish body. When it is angry or upset, these circles glow like neon signs. This warns its enemies that it is poisonous and should not be eaten.

Disappearing Trick
Many animals are so beautifully camouflaged that they are very hard to see. Some can even change the color of their skin!

Blue swirls

These mandarin fish may look bright, but when they are swimming in gently rippling, sunlit seas, their swirls of color make them almost invisible!

Flatfish are very tasty, so lots of bigger fish like to eat them. This flounder can make patterns on its skin that match the sand and rocks around it.

Seaweed

Can you spot this pipefish? Its long, thin body looks just like the seaweed it lives in.

What Animal?
These animals are pretending to be rocks and weeds. If they don't look like a tasty meal, then hungry animals might leave them alone.

Eye

Stonefish have bumpy bodies that look like stones. But if you step on one, it will sting you!

Mouth

This crab has tucked its legs under its shell and is sitting very still. With luck, no fish, octopuses, or seals will find it.

WHALES

Millions of years ago whales used to walk. Since then, they have changed a lot. They have grown bigger, their back legs have disappeared, and their front legs have turned into flippers. They can't live on land anymore, but they are still mammals. This means they breathe air and feed their babies milk.

There are two kinds of whales: toothed whales and toothless whales, called baleen whales.

Krill for Lunch
Baleen whales eat krill, a kind of plankton. A large whale can eat two tons a day – half the weight of an elephant!

The whale moves its tail up and down to push itself through the sea. The fastest whale is the sei. It can reach speeds of 30 miles an hour (48 km/h), ten times faster than a person can swim.

No animal has longer flippers than a humpback whale – they are about five yards (5 m) long, almost as tall as a giraffe!

Whales have thick fat, called blubber.

There She Blows!
When a whale surfaces to fill its lungs with fresh air, warm air escapes from its blowhole. This escaping air mists up, just like your breath on a cold day, and forms a tall spout called a blow.

Deep throat grooves let a baleen whale's mouth stretch to hold vast amounts of seawater and krill.

Jumping for Joy
Humpback whales love to leap right out of the water. This is called breaching. When they crash back into the ocean, they make a huge splash!

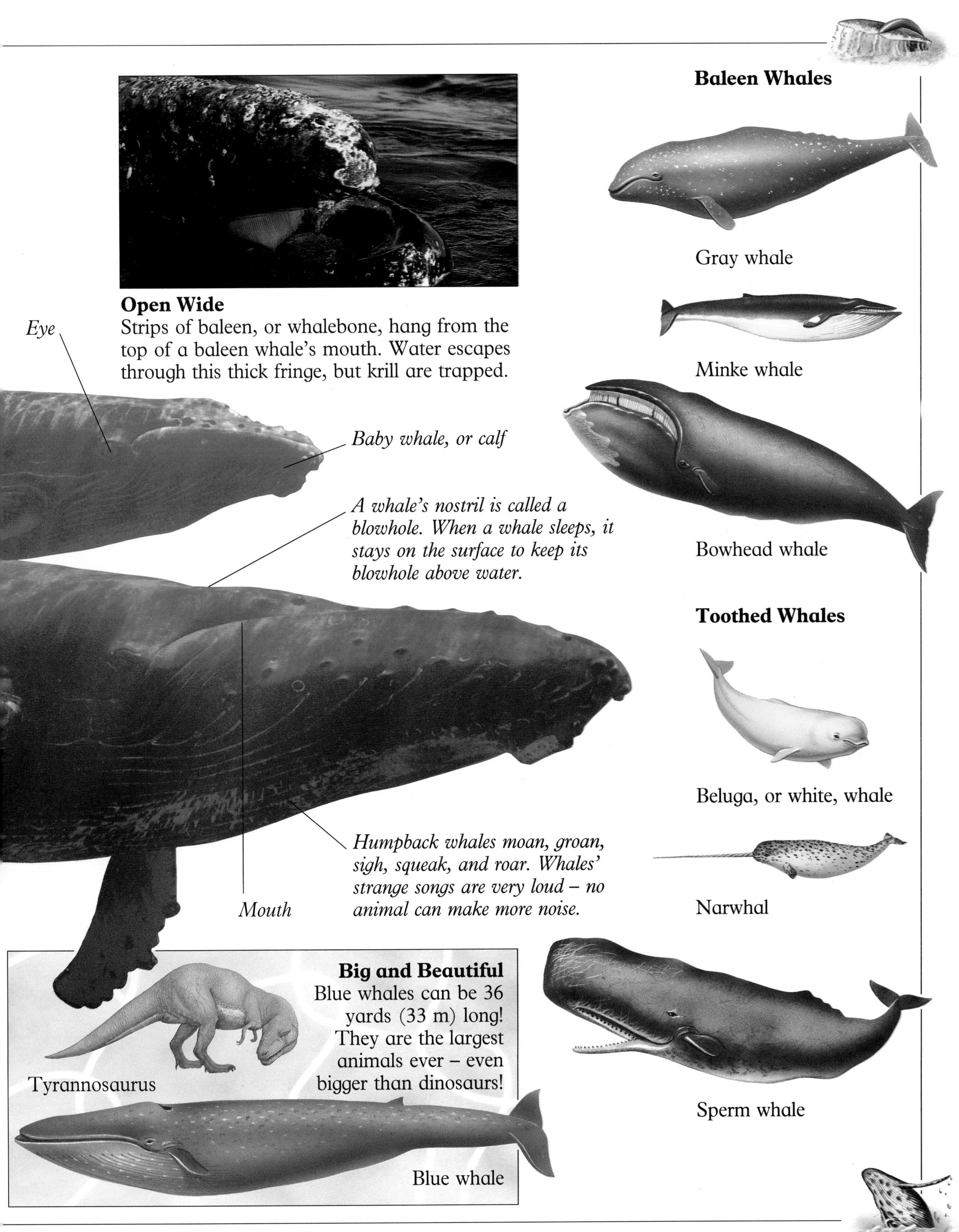

Open Wide
Strips of baleen, or whalebone, hang from the top of a baleen whale's mouth. Water escapes through this thick fringe, but krill are trapped.

Big and Beautiful
Blue whales can be 36 yards (33 m) long! They are the largest animals ever – even bigger than dinosaurs!

DOLPHINS

Dolphins are small, toothed whales. Some people think that these smooth-skinned mammals are highly intelligent. They learn quickly and seem to talk to one another with whistles, clicks, and grunts. Since ancient times, there has been a special friendship between humans and these playful animals. There are many stories of dolphins saving drowning sailors.

Large, curved dorsal fin

Brain

Melon

Dolphins do not need to drink. They get all the water they need from the fish and squid they eat.

Built-in Sonar
In the dark, dolphins use sound to find their food and to figure out where they are. The melon, a fatty area in the head, sends out clicks that bounce back when they hit something. These echoes tell dolphins where an object is.

Sonar clicks

The shape of a dolphin's mouth makes it look as if it is always smiling.

Toothy Grin
Bottlenose dolphins have more than 100 teeth, each about one-third of an inch (8 mm) long.

Hitching a Ride
Just in front of a speeding ship, there is a wave. Dolphins love to swim in this spot. Like surfers, if they catch the wave in the right way, it will push them through the sea.

Blowhole

Most dolphins have a long snout, called a beak.

Family Life
Dolphins live in family groups called herds. Baby dolphins often stay with their mothers for many years. They learn how to catch fish, signal to each other, and escape from sharks by copying other members of their herd.

Dolphins, like all toothed whales, have only one blowhole. Baleen whales have two nostrils.

Bottlenose dolphins are very playful. They love to leap out of the water.

Tail fluke

Just as you have your own personal name and voice, every dolphin has its own whistle, which other dolphins recognize.

The Dolphin Family

Harbor porpoise

Bottlenose dolphin

Striped dolphin

Common dolphin

Risso's dolphin

Killer whale (male)

The Biggest Dolphin
Killer whales are fierce dolphins that can grow to be 30 feet (9 m) long. This one is trying to grab a sea lion.

SEALS

Seals are warm-blooded mammals, which means that they can make their own heat. Because they often swim in very cold water, they need to be able to keep this heat inside their bodies. Seals can't put on a warm coat like you do to keep warm. Instead, they are covered with short hair or fur and have a layer of fat, called blubber. Sometimes, when seals swim in warmer water, they get too hot and have to fan their flippers in the air to cool down.

Ball of Fluff
Baby seals, or pups, are born on land. For a few weeks, many of them have white, fluffy fur.

All seals have good hearing, but only sea lions and fur seals have ear flaps. All that can be seen of this harbor seal's ears are two tiny holes.

When a seal dives under water, it closes its nose, mouth, and ears!

These whiskers, which are 40 times thicker than human hair, can sense movement in the water. This helps the seal find fish, shellfish, squid, and octopus to eat.

Pile Up!
When walruses climb up onto beaches, they often lie right on top of each other to keep warm!

Flip Flop
Sea lions, fur seals, and walruses use their front flippers to sit up straight, and their back flippers can turn forward. This means that they can walk, and even run, on dry land. Hair seals can only slide around on their bellies when they leave the sea.

The Seal Family
There are three different groups of seals:

Gray seal

Hair seals

Elephant seal

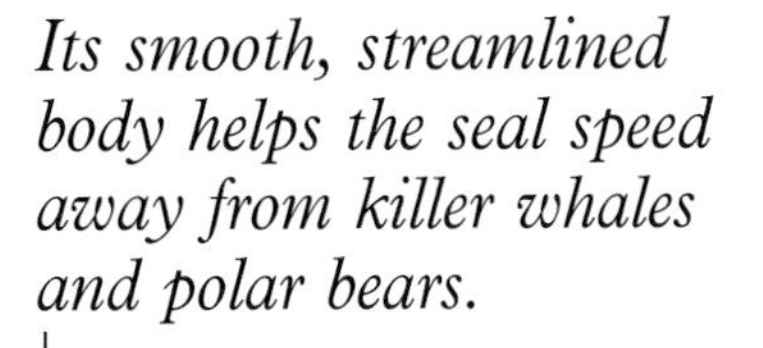

Its smooth, streamlined body helps the seal speed away from killer whales and polar bears.

Eared seals

Northern fur seal

Blubber is about four inches (10 cm) thick.

Seal-eating Seals
Leopard seals are fierce. They leap out of the sea and thump onto the ice to grab penguins or other seals.

California sea lion

This seal is less than two yards (2 m) long. An elephant seal can measure nearly six yards (6 m) from nose to back flippers, which is more than the height of a giraffe!

Walruses

Male walruses fight each other with their big teeth.

Hair seals, like this harbor or common seal, speed through the ocean by moving their back flippers up and down.

SEABIRDS

Seabirds live all over the world and lead very different lives. Penguins swim under ice, gulls swoop over stormy waves, and albatrosses soar high in the sky above tropical islands. But all seabirds have one thing in common: they have to come ashore in the summer to lay their eggs, often on cliffs, in crowded, very noisy "bird-cities."

Balloon Bird!
This male frigate bird is inflating his red throat to attract a female.

Puffin eggs, actual size

Male puffins fight with their colorful beaks.

Eye

Seabirds drink salty seawater. The salt then drips out of their nostrils and back into the ocean.

Fish can't see a puffin that is flying low over the sea. This is because a puffin's white chest feathers are the color of the sky.

Bird Watching
Petrels, like many birds, quickly join in when they see another bird feeding.

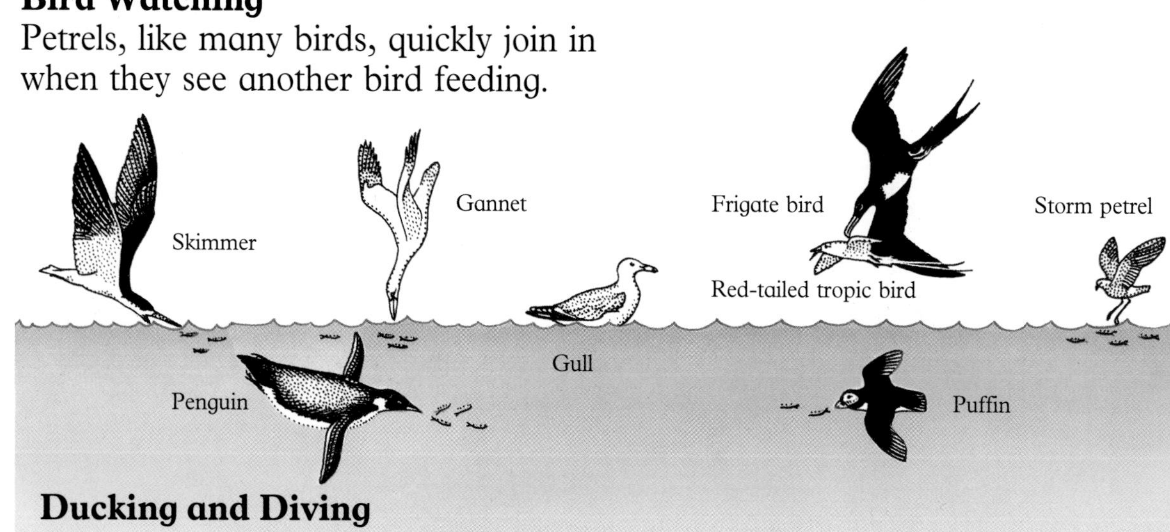

Ducking and Diving
Seabirds catch fish and plankton in many different ways. Gannets dive-bomb into the sea and frigate birds often steal another bird's food!

Each year, puffins lay one white egg in a grass-lined burrow at the top of a cliff. They dig a two-yard-long (2 m) hole with their claws.

Webbed foot

Bird Parade

Seabirds come in many different sizes. The girl holding this ruler is four feet (1.25 m) tall, but she is only a bit taller than the biggest penguin.

Up, Up, and Away

The white wings of a wandering albatross can measure more than 11 feet (3.3 m) from tip to tip. Keeping their wings still, they can glide around the skies, and travel more than 500 miles (800 km) a day. For the first four years of their lives, they stay in the air most of the time.

Emperor penguin

Frigate bird

Blue-footed booby

Herring gull

Red-tailed tropic bird

Tufted puffin

Most seabirds' feathers are covered in a special oil that makes them waterproof.

Flight feather

Short tail

Air that is trapped under the feathers keeps seabirds warm.

Puffins are not strong fliers – they have small wings and heavy bodies. But their wings make great flippers when they are under water.

Far Too Fat!

Gannet chicks eat so many fish that they get too fat to fly away with their parents. Instead, they leap off the cliff where they were born, crash into the sea, and start to swim. Only after they have swum hundreds of miles are they slim enough to stagger into the air.

SEA TURTLES

A big, bony shell covers all of a sea turtle's body except its head and legs. Its flat, smooth shape lets a turtle surge through the water at speeds as high as 40 miles per hour (65 km/h) if it has to. This incredible suit of armor also helps protect the sea turtle from being eaten by sharks and killer whales.

Overlapping scales

A sea turtle's legs and head are too big to be pulled inside its shell.

What's Inside a Turtle?
The bottom half of this turtle's shell has been removed so that you can see its skeleton.

Claw

Tail

Skull

Neck

Rib joined to shell

Backbone joined to shell

The scales lie on top of this bony shell.

Like all reptiles, sea turtles are cold-blooded. This means that their body temperature can go up and down with the heat. If the sea is warm they are warm, if the sea is cold, they are cold.

Its front flippers work together like the oars of a rowboat to push the turtle through the water.

Baby turtle

Egg

Sea Turtles Hatching
A turtle lays her eggs on a beach. She digs a hole with her strong back flippers, lays more than 100 eggs, then covers them with sand to keep them warm. When the babies hatch, they scramble quickly down to the sea to avoid being eaten by crabs and birds.

Large and Leathery

Leatherback turtles can be two yards (2 m) long and have leathery skin instead of bony scales on top of their shells.

Claw

Turtles do not have teeth – they have a powerful bony beak.

Eye

Nostril

This hawksbill turtle likes to eat crabs and squid. Green turtles are different – their favorite foods are seaweed and sea grass.

Sea Turtle Types

Hawksbill turtle

Green turtle

Pacific ridley turtle

Loggerhead turtle

Leatherback turtle

Sea Turtles in Danger

The large scales that cover a sea turtle's shell are made of keratin, like your fingernails. They are often called tortoiseshell. People used to kill turtles to make things from these beautiful scales. This is not legal anymore because there are so few turtles left.

All these objects are made from tortoiseshell.

LONG JOURNEYS

Many animals spend their whole lives in one place. Others travel across oceans in search of warmth, food, or a place to give birth to their young. These regular journeys are called migrations. Nobody understands how animals know the way to places they have never visited before. Perhaps they look at the Sun, Moon, and stars ... or maybe they have a built-in compass!

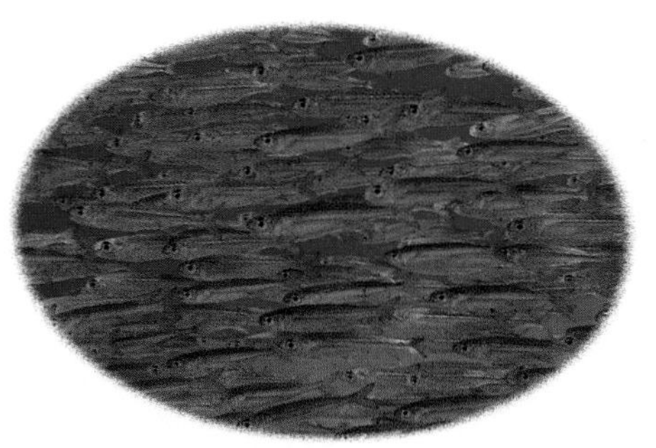

All Together
Herring have a regular travel timetable. They migrate from their winter home in deeper, warmer waters, through their feeding areas, and on to their summer breeding grounds.

Long, thin "claw"

Keeping a Lookout
Every October, gray whales that live near the North Pole start to swim south. When they reach the warm waters off the coast of Mexico, they give birth. In March, they head back north. During this trip of up to 12,000 miles (19,300 km), they often peer out of the sea. Maybe they are trying to see where they are.

Up to 60 lobsters march in a line. They keep together by touching one another with their feelers.

Sometimes spiny lobsters hook their front claws around the lobster in front of them. This keeps them from being washed away by strong currents.

Follow the Leader
The spiny lobsters that live off the coast of Florida start to head south as soon as the first storms of winter arrive. After they have traveled more than 60 miles (100 km), they reach calmer, warmer seas. There they can lay their eggs. They may not migrate as far as whales, birds, and fish, but lobsters have to walk across the ocean floor. They cannot swim or fly!

Uphill Struggle
To lay their eggs, salmon leave the ocean and swim back up the river in which they were born. Nothing stops them – they even leap up steep waterfalls.

Summer-loving Bird
Arctic terns spend the summer in and around the Arctic. When it gets colder and there is less food, they fly south to the Antarctic and spend summer there, too. In a lifetime, they travel about the same distance as a trip to the Moon and back!

Eye

Tough, spiny shell

Lobsters have four pairs of legs and can walk faster than humans can swim.

DOWN IN THE DEPTHS

Light cannot reach the very bottom of the ocean, so it is always pitch black and as cold as the inside of a refrigerator! But even in this unlikely place, fantastic fish survive. If you could dive down this far, you would see tiny pinpricks of light that are actually glowing deep-sea fish. These patterns of light help them to be seen in the dark, just like clothing reflectors people wear at night.

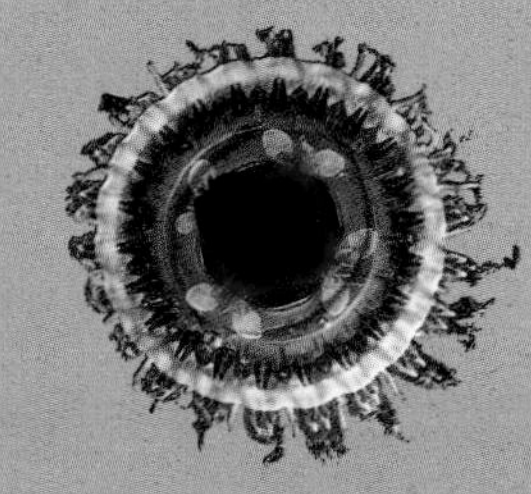

Drifting By
Beautiful jellyfish float silently around the bottom of the sea.

Eye

Silvery scales

These stripes glow in the dark.

Eye Spy
This hatchet fish lives in the twilight zone, where there is still some light left. Its large eyes point up toward the light at the surface.

Like most deep-sea fish, this female angler fish is less than four inches (10 cm) long.

Male

Eye

This small tail fin means that the angler fish can swim only very slowly.

Sitting on Slime
Many animals live on the thick slime that covers the bottom of the sea. They eat the remains of dead plants and animals that sink down from above.

Sea cucumber

Sea spider

Venus flower basket, a type of sea sponge

Tripod fish

650 feet (200 m)
Twilight zone
3,300 feet (1,000 m)
20,000 feet (6,100 m) Trench

Water Colors 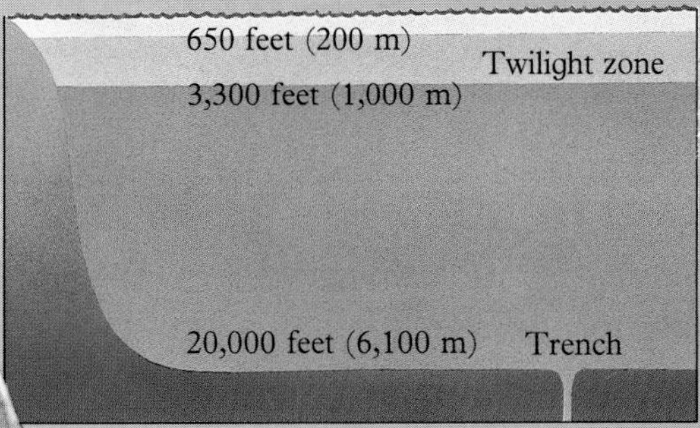
Most of the sea is about 20,000 feet (6,100 m) deep. Light can't travel very far down, so only the top part is brightly lit. Below 3,300 feet (1,000 m), it is totally dark.

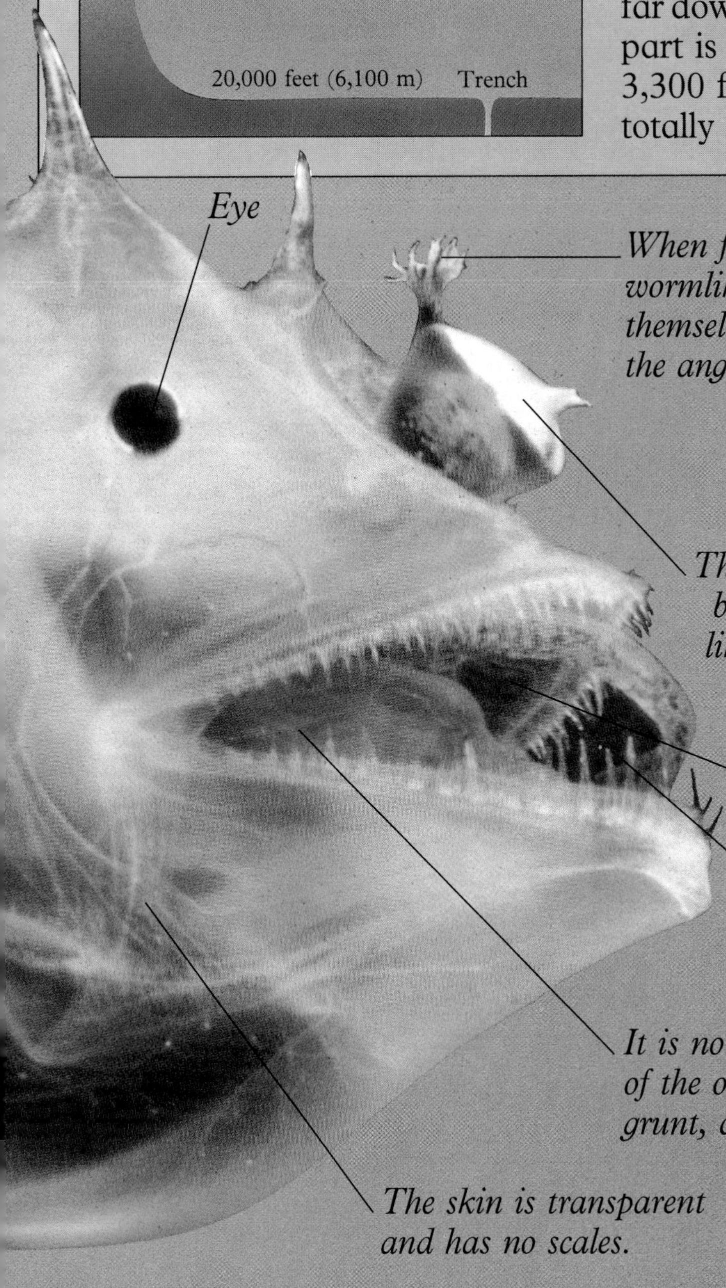

Eye

When fish try to eat this wormlike bait, they find themselves being eaten by the angler fish instead.

This shining area can be turned on and off like a flashlight. It is used to scare bigger fish away.

Big mouth

Jagged teeth

It is not silent at the bottom of the ocean. Many fish can grunt, click, and even whistle.

The skin is transparent and has no scales.

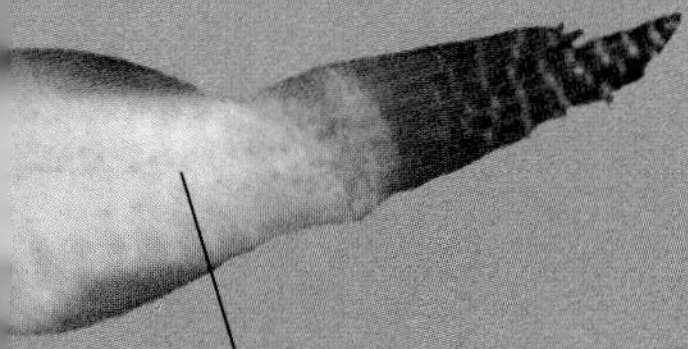

This male angler fish has stuck itself onto a big female.

Scarlet Swimmers
Food is scarce down in the depths, but bright red prawns are the favorite food of many deep-sea swimmers.

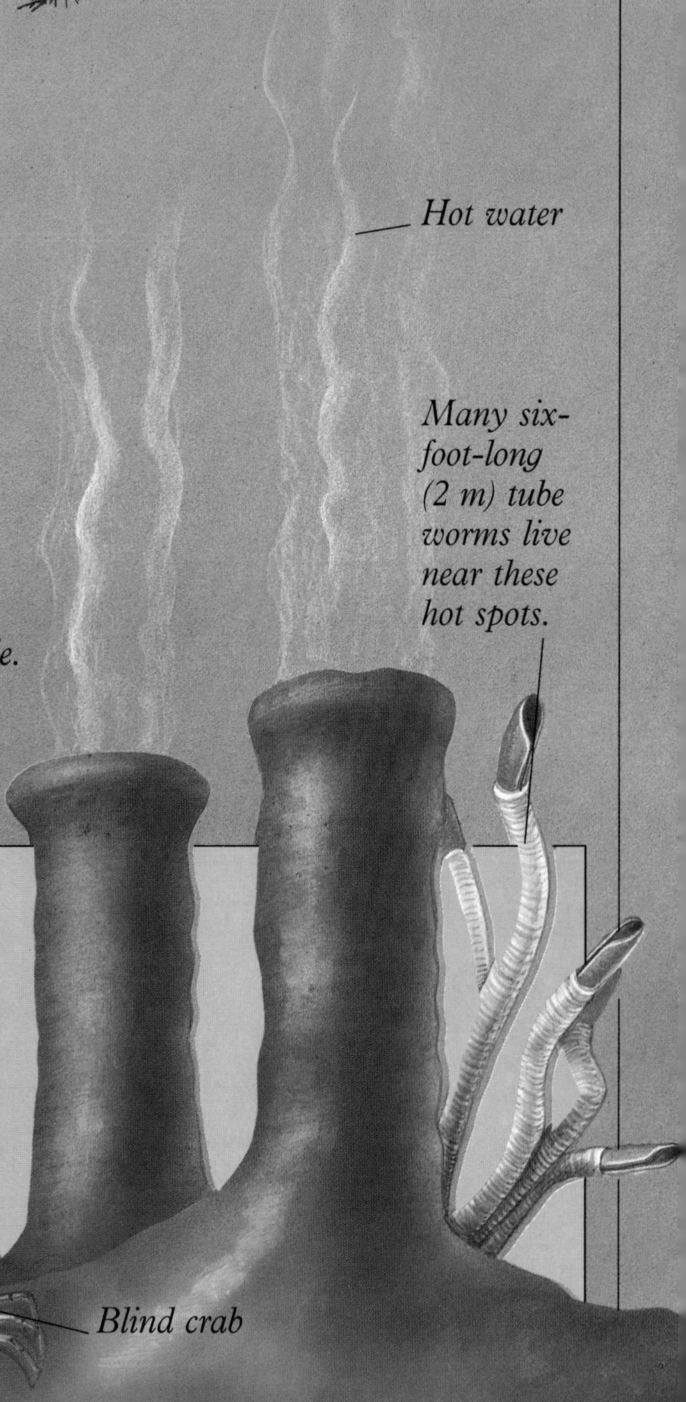

Hot water

Many six-foot-long (2 m) tube worms live near these hot spots.

Hot Water Fountains
In some spots, boiling water bursts through the slime, like an underwater volcano. The sulphur in this water piles up and forms a chimney. If the chimney gets too tall, it collapses and the hot water escapes, killing the animals that live nearby.

Blind crab

GLOSSARY

Backbone A line of bones found inside the back of reptiles, birds, mammals, and some fish.

Baleen Horny material that hangs inside the mouth of some whales and strains their food from the water. It is also called whalebone.

Blowhole The nose, or nostril, of a whale, situated on top of its head.

Blubber A thick layer of fat that keeps whales, seals, and walruses warm.

Bone A hard, chalky material inside the body. It forms skeletons.

Camouflage The color patterns or body shapes that help hide an animal from its enemies.

Cartilage A rubbery material that forms the skeleton of sharks and rays.

Cold-blooded The temperature of a cold-blooded animal goes up and down with the temperature of its surroundings.

Coral Small sea animals that can be hard or soft. Hard corals build chalky homes around their bodies.

Egg An unborn baby and its food supply, which is usually protected by a shell.

Fin A flat piece of skin and bone attached to the side or back of a fish. Fish use fins to help them move through the water.

Fin ray A small, stiff rod that supports a fin.

Flipper The paddle-shaped legs of whales, dolphins, seals, or turtles.

Gills Parts of the body that absorb oxygen from the water and let animals breathe under water.

Joint The point where bones meet. Knees and elbows are joints.

Larvae An early stage of life for animals such as fish, crabs, and shellfish. Larvae look different from adults.

Lung A part of the body that is used for breathing air. Mammals have lungs.

Mammal Animals that breathe air, have warm bodies, and give birth to live young that they feed on milk. People are mammals.

Migration The regular movement of animals from one place to another to find food or escape cold weather.

Mineral A chemical, such as salt or iron, that all living things need to have in order to survive.

Mollusk Shellfish, which are related to land-living animals such as snails and slugs, are mollusks.

Molting Shedding an old, and growing a new coat of skin, fur, or feathers.

Ocean A huge area of saltwater that is bigger and deeper than a sea.

Oxygen A gas needed for breathing. It is found in the air and in water.

Plankton Tiny plants and animals that float in the sea and are eaten by many other animals.

Reef A long wall of coral skeletons that grows in shallow, warm seas, near to the shore.

Reptile A cold-blooded animal that is covered in scales. Reptiles breathe air and most hatch from leathery eggs.

Salt The most common mineral in the ocean. It makes the water taste salty.

Scales Small, flat, hard plates that cover the skin of some animals.

Shell The hard outer covering of shellfish, crabs, and lobsters.

Skeleton The hard parts of an animal that support its body. The skeleton can be made of either cartilage or bone.

Snout Part of the head of an animal. A snout is often long and has the mouth and nostrils at the end of it.

Spawning The squirting of millions of tiny soft-shelled eggs into the sea.

Species A sort of animal. Any male and female of the same species can breed together. A male killer whale and female sei whale cannot have a baby – they are different species.

Streamlined Sea animals that have rounded bodies are streamlined. The smooth shape of their body helps them move through the water faster.

Tentacle A long arm, or feeler, of an animal.

Tide The rise and fall of the sea level. This is caused by the pull of the Sun and Moon on the water.

Trenches Wide, very deep ditches that are found on the bottom of the ocean.

Acknowledgements
Photography: John Edwards and James Stevenson. **Illustrations:** Annabel Milne and Roy Flooks. **Models:** Truly Scrumptious Child Model Agency.
Thanks to: Lee Marshfield at Portsmouth Sealife Centre and Dr. Potts at Plymouth Marine Association.

Picture credits
Ardea: François Gohier 34b, 35t, 37, 44b, Mike Osmond/Auscape International 4/5c, 34/35, Ron & Valerie Taylor 26; **Biofotos:** Heather Angel 38t, 47; **Bruce Coleman:** Francisco Erize 39t, Dr. Inigo Everson 34t, Jeff Foott Productions 10tl, 36/37, Dieter & Mary Plage 7, Hans Reinhard 38b; **Robert Harding Picture Library:** Schuster/Kosel 30bl; **Dianne R. Hughes,** (Biological Sciences, Macquarie University, NSW, Australia) 27tc; **Frank Lane Picture Agency:** H. D. Brandt 45tr; **NHPA:** Anthony Bannister 43, J. Carmichael 32bl, B. Jones & M. Shimlock 2b, 31b, Ashod Papaziar 29, John Shaw 45tl, Roy Waller 1, 10tr, Bill Wood 24clb; **Oxford Scientific Films:** Doug Allen 39b, Kathie Atkinson 14cr, Laurence Gould 30c, Mark Hamblin 5tr, 40/41, back cover br, Frank Huber 5crb, 45b, T. S. McCann 41t, Ben Osborne 40b, Peter Parks 4c & clb, 12t & r, 12/13t & b, 46tr, Frank Schneidermeyer 36t, Wisniewski/Okapia 40t; **Pictor** endpapers; **Planet Earth Pictures:** Leo Collier 19b, Mark Conlin 3c, 11tr, Peter David 46/47, Walter Deas 32br, Georgette Douwma 25cra, A. Kerstich 20cl, Kenneth Lucas 10b, 21tc, 23b, Doug Perrine 4b, 14/15, Christian Petron 15br, Peter Scoones 24bl, 25bl, 33cb, 44t, Herwath Voigtmann 28, James D. Watt 5b, 24c, 36b, 38/39, Margaret Welby 41b, Norbert Wu front cover clb, 5cb, 20tr, 46l; **Science Photo Library:** Dr. Gene Feldman/NASA/GSFC 13t, Tom Van Sant/Geosphere Project, Santa Monica 6tl; **Survival Anglia:** Annie Price 42.

t – **top** l – **left** b – **bottom** r – **right** c – **center** a – **above** cb – **center below** clb – **center left below** crb – **center right below**

INDEX